संख्या को कथा

THE NUMBER STORY

SMALL BOOK ONE

ENGLISH - NEPALI

Numbers Teach Children
Their Number Names

written and illustrated by

MISS ANNA

Early Reader Edition of *The Number Story 1*
Bronze Medal Winner, 2016 Wishing Shelf Book Award

LUMPY PUBLISHING

Library of Congress Control Number: 2018902040

Names: Miss Anna, author.
Title: Number story : numbers teach children their number names / Miss Anna.
Description: Portland, OR: Lumpy Publishing, 2018.
Identifiers: ISBN 978-1-945977-8| LCCN 2018902040
Summary: The pictures and rhymes present stories which introduce numbers 0-10.
Subjects: LCSH Numeration—English--Nepali--Pictorial works--Juvenile literature. | BISAC JUVENILE NONFICTION /
Languages: English--Nepali
Classification: LCC QA141.3 .M57 2018 | DDC 513—dc23

Publisher: Lumpy Publishing
Website: www.missannabooks.com
Email: missanna@missannabooks.com

Paperback: ISBN 978-1-945977-89-3
Printed in the U.S.A.   1 3 5 7 9 10 8 6 4 2

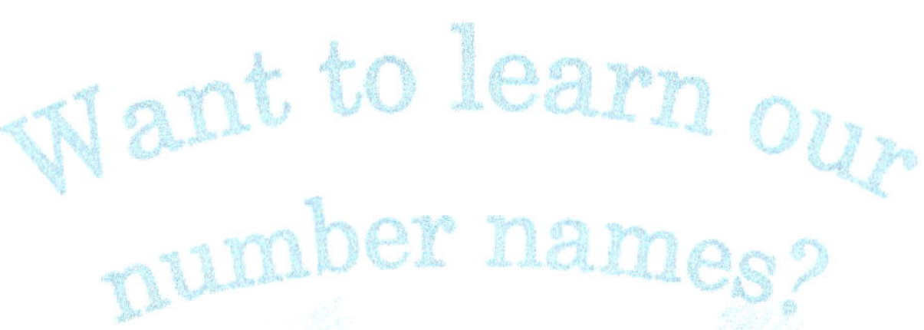

हाम्रा संख्याहरुको नामहरू सिक्न चाहनुहुन्छ?

It is very easy and a lot of fun!

यो धेरै सजिलो र धेरै रमाइलो छ!

Say-along our little jingle

हामी संग हाम्रो सानो कथा गाउनुहोस्!

starting from Number One!

हामी संख्या एक बाट सुरू गरौं!

1

ONE looks like my one finger.

१ ☆ एक

एक मेरो एक औंला जस्तो देखिन्छ।

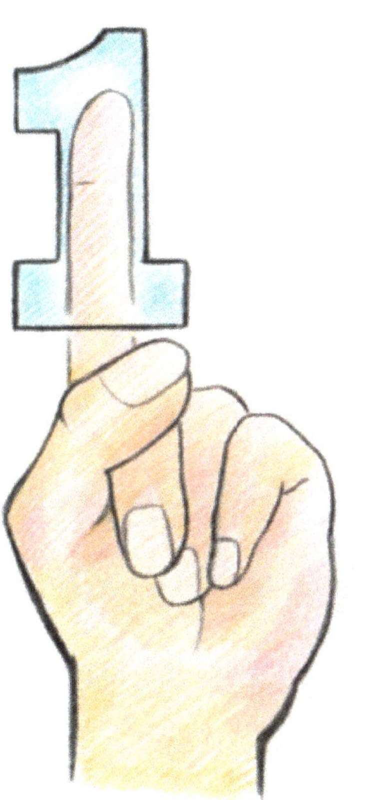

ONE!
एक!

2

TWO trails a tail.

२ ☆ दुई

दुई पुच्छरलाई पछ्याउछ।

A TAIL! एक पुच्छर!

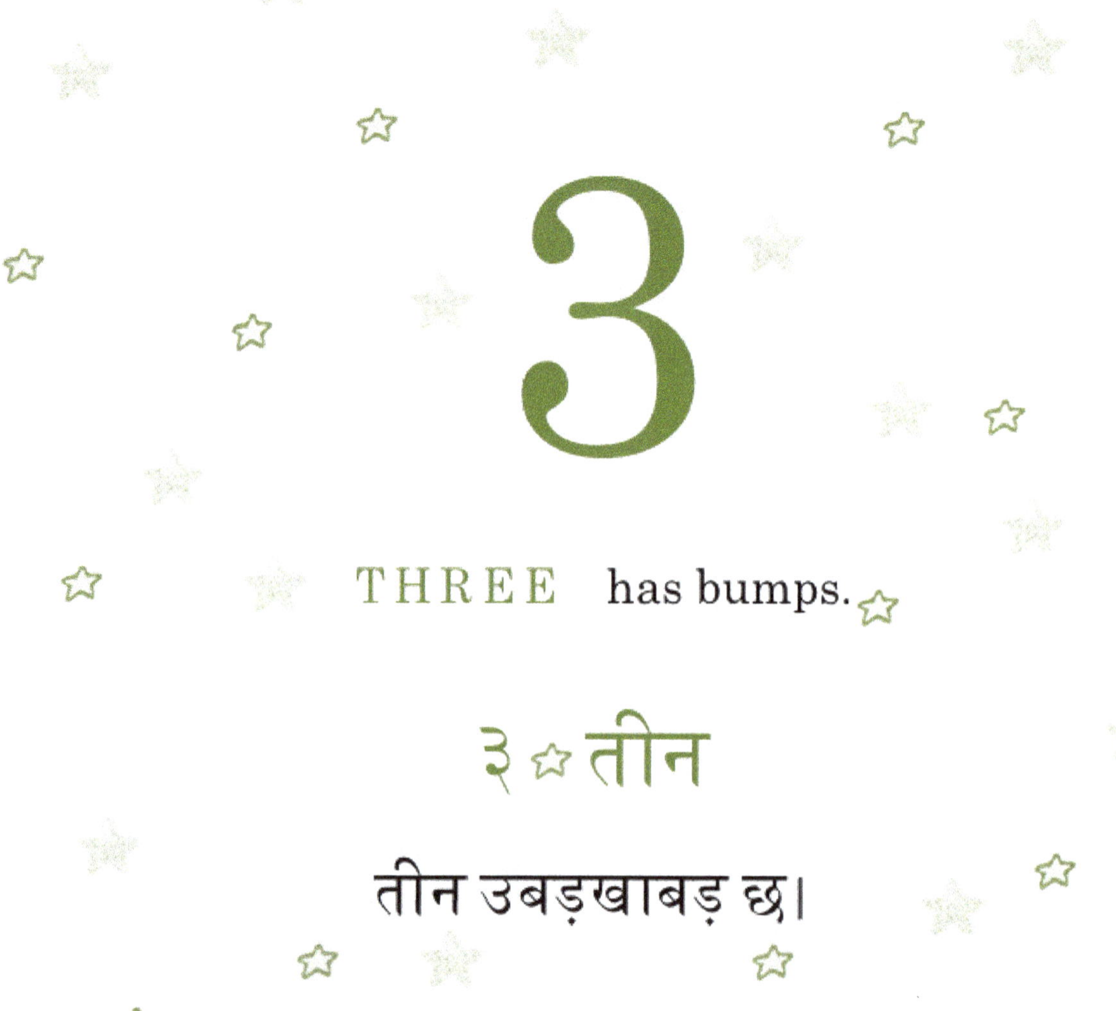

3

THREE   has bumps.

३ ☆ तीन

तीन उबड़खाबड़ छ।

वक्रहरू हेर्नुहोस्!
हरित पहाडहरू हेर्नुहोस्!

4

FOUR carries a sail.

४ ☆ चार

चारले पाल राखेको छ।

A SAIL!
एक पाल!
पाल भएको एउटा डुंगा!

5

FIVE   is a racing track.

५☆ पाँच

पाँच रेस गर्ने बाटो हो।

VROOM!
वरूम

SIX   curves like a snail.

छ एक चिप्लेकिरो जस्तो बांगो तेर्सो चल्छ।

A SNAIL!   एक चिप्लेकिरो!

7

SEVEN has a sharp angle.

७ ✦ सात

सातमा एक तीव्र कोण छ।

BE CAREFUL!
सावधान हुनुहोस!

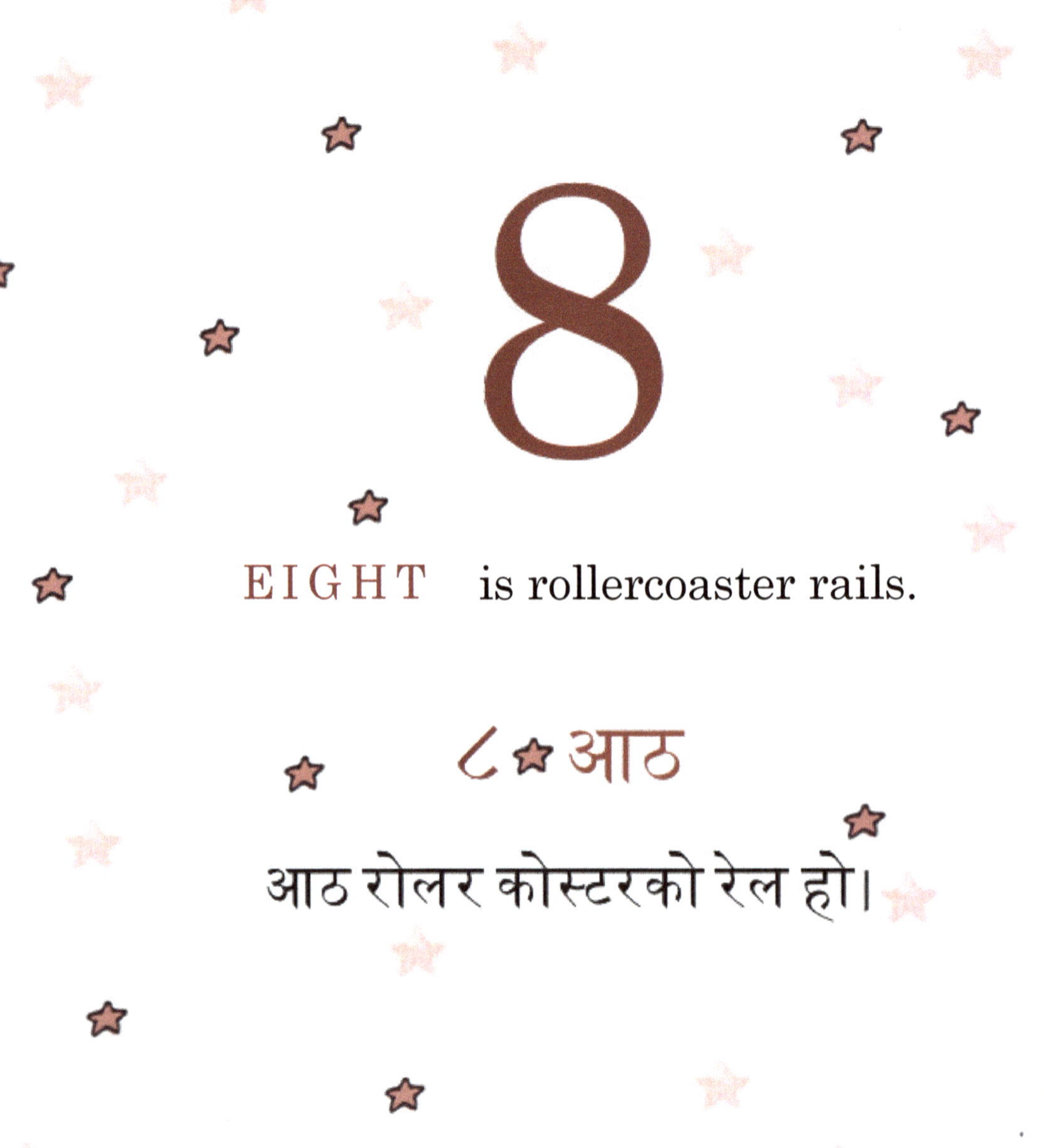

8
EIGHT   is rollercoaster rails.
८ ★ आठ
आठ रोलर कोस्टरको रेल हो।

इप्पी!
YIPPEE!

NINE   is a bubble on a stick.

९ ❈ नौ

नौ एक लठ्ठीमा एउटा बुलबुला हो।

A BUBBLE!   एउटा बुलबुले!

10

TEN   is an eye of a whale.

१०☆दस

दस एक व्हेलको एउटा आँखा हो।

HELLO!
हेलो!

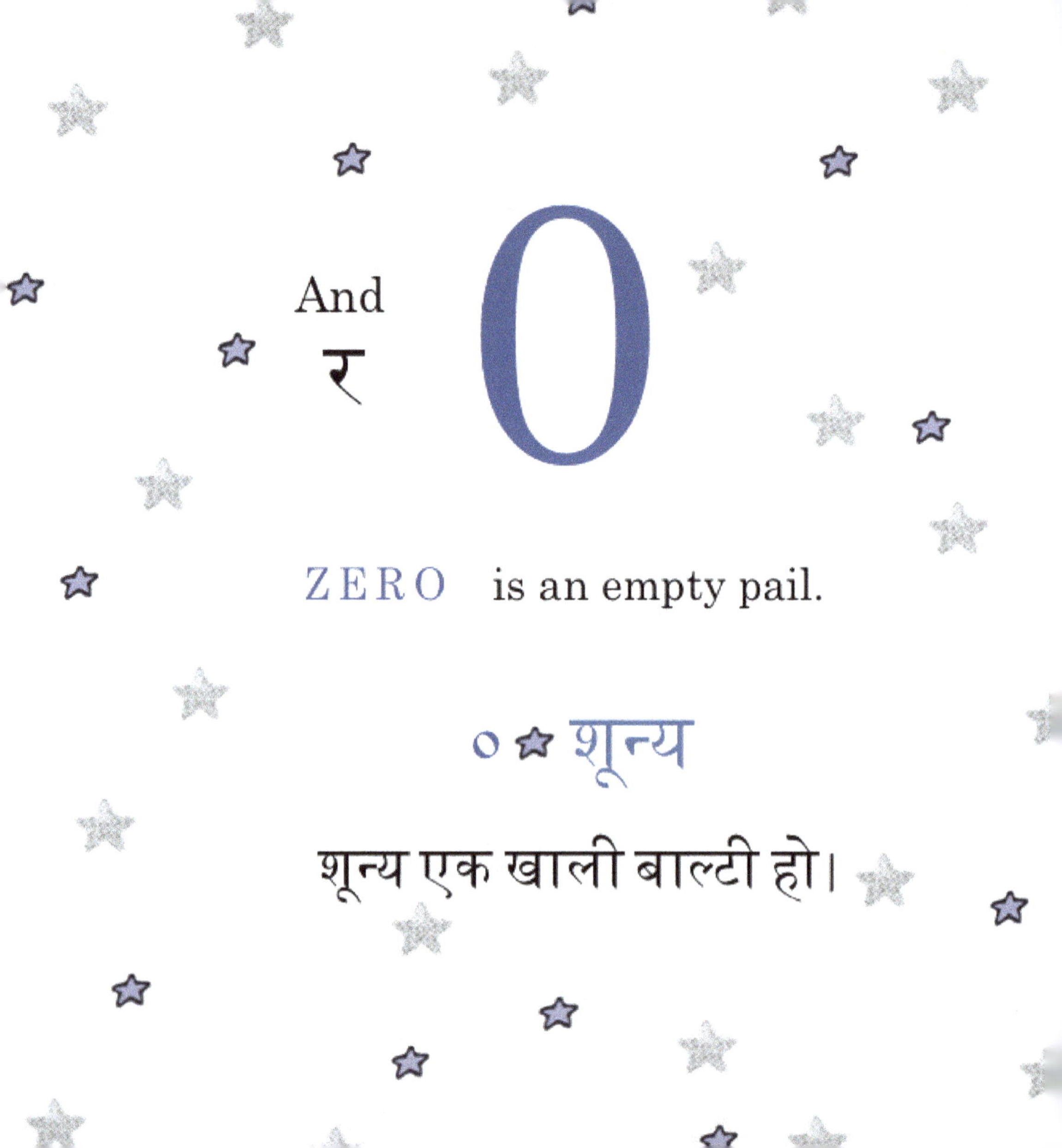

And

र

0

ZERO is an empty pail.

० ☆ शून्य

शून्य एक खाली बाल्टी हो।

IT'S EMPTY!
यो खाली छ!

Thank you for playing with us today.

We had a lot of fun too!

आज हाम्रो साथ खेल्नको लागी धन्यबाद।

हामीले धेरै रमाइलो गर्‍यौं!

We are your Number friends,
Zero to Ten,
Who will be here for you~

हामी तपाईको संख्या साथीहरु छौं
शून्य देखि दस सम्म।
हामी सधैं तपाईको लागि यहाँ हुनेछौं।

Bye-bye now!
See you again soon.
अहिलेको लागि बिदा!
चाडै फेरी भेटौला।

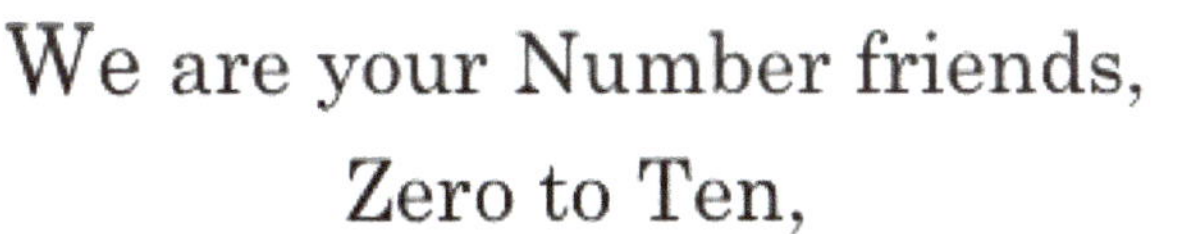

The Numbers are *SINGING* too!

To sing-a-long, look for Miss Anna Number Story
at your favorite music store like iTUNES.

MP3

Numbers 0-10
IDENTIFYING
& COUNTING

Numbers 11-20
& Ordinals

first, second, third...

Numbers 0-100
& Place Values

ones, tens, hundreds...

About Clocks
& Telling Time

hours, minutes, second

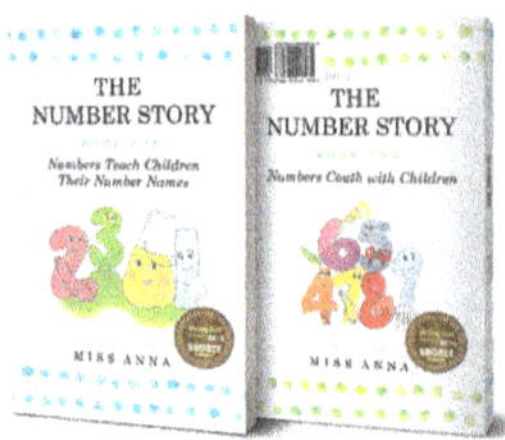

Number Story 1 & 2
isbn: 978-0-996216-48-7

Number Story 3 & 4
isbn: 978-1-945977-01-5

Number Story 5 & 6
isbn: 978-1-945977-06-0

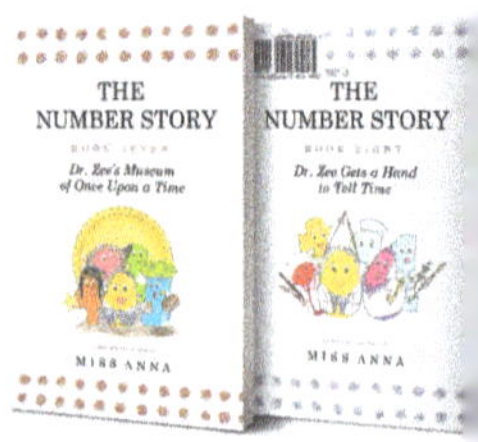

Number Story 7 & 8
isbn: 978-1-949320-40-

For more Miss Anna books to love,
visit us at

www.missannabooks.com

Numbers are working hard all over the world!
Come Travel the World with Us!